The Beauty of the Palouse
A geologic and poetic story

Written and Illustrated by the 2016-17 Fourth Grade Crew at Palouse Prairie Charter School

Felicia Fiorillo
Jamison Griffin
Eric Kozlowski-McLeod
Ruby Miller
Claudia Monroe

Carly Royals
Sunflower Stanton
Carmen Stevens
Camas Stone
Jackson Taylor

I0393417

Fourth grade crew enjoying a snowy day on the Palouse.

The fourth grade crew at Palouse Prairie Charter School wrote this book as the final product of a semester-long study about the geologic formation of the Palouse region around Moscow, Idaho. They demonstrated genuine responsibility and grit by taking on the challenge of being the first to write this story. In order to complete this project these students studied and practiced the arts of scientific writing, poetic writing, and linoleum block printmaking. This book exemplifies each student's dedication to scientific accuracy, creativity and craftsmanship. We hope you enjoy the scientific and poetic stories, appreciate the artistic illustrations and learn something new about our beautiful Palouse!

—*Renée Hill, Fourth Grade Teacher*

The fourth grade crew would like to thank our experts for
so generously sharing their time, knowledge and encouragement.
Thank you each for helping make this project possible, beautiful and fun!

Dennis Feeney, Geologist, Idaho Geological Survey (dmfeeney@uidaho.edu)
Krista Brand, Artist, www.kristabrand.com
Hannah Kroese, Graphic Designer, http://hannahkcrawford.com
Stacy Boe Miller, Poet
Terry Pierce, Children's Author, http://terrypiercebooks.com

ISBN: 978-1-5429-3325-4

Available from Amazon.com and other retail outlets

THE rolling hills and mountain peaks of the Palouse you see today haven't always been here or looked this way. Have you ever wondered how the Palouse was formed? We wrote this book to tell two wonderful geologic stories about the Palouse. One is a scientific explanation and the other a poetic tale.

DID you know that 1.5 billion years ago the Palouse was under the sea? The Palouse region was just a small bit of seabed where sand and sediment were being deposited and resting for billions of years. When you're looking out over the Palouse region now, just imagine that long ago this land was entirely under the ocean.

I am the past Palouse

a pile of rough silt
under the sea, brother of Earth
alone I inch to go up to land
the swooshing is maddening,
all I want is to go up.

I am a grain of sand that will
become a magnificent
land.

THE Earth's surface is made of solid rock called the lithosphere. It is broken up into a puzzle of pieces called plates. Underneath is a partially melted layer of rock called the asthenosphere. It acts like a banana peel that the plates slip and slide on. Plates move by colliding, moving apart, and slipping past each other. That is the process of plate tectonics. Plate movement causes rock to move, shift, crack, break, and fold, creating earthquakes, volcanoes and mountains. Many millions of years ago plate tectonics created high peaks and low valleys on the Palouse. This region used to be filled with mountains much larger than what we see today.

I sound like I'm crashing

growling, and banging.
I do powerful things.
I transform, converge,
and diverge.
I fault, I fold
I slip, I slide
I crush, I create.
I make peaks and valleys.
But slowly.

Without me
none of the Palouse you see today
would be here.

I am plate tectonics.

QUARTZITE was formed from the sand resting and accumulating on the bottom of the seafloor back when the Palouse was under the ocean. All of the movement of the Earth's plates resulted in layers of rock being piled up on top of the sand. As the sand was compressed and compacted by thousands of feet of rock on top of it, a sedimentary rock called sandstone was formed. Then, over millions of years, as the sandstone was exposed to extreme heat and pressure deep inside the Earth, a metamorphic rock called quartzite was formed. This giant lump of quartzite waited for millions of years to become the largest peak on the Palouse.

I was squished

squeezed
mashed
squashed and scrunched,
pounded
crushed and flattened.
I was baked
burnt, scorched
toasted and fried.

I am now a
towering
colossal
monstrous mass.

I am quartzite.

WITH the folding and faulting of plate tectonics, this giant mass of quartzite was uplifted and shaped into today's largest formation on the Palouse—Kamiak Butte. For millions of years, Kamiak Butte has been baked by the sun and cooled and sculpted by the wind and rain. It now stands 3,641 feet above sea level. Although it is taller than the world's tallest building, it stands only about 1,100 feet above the Palouse Hills. It may be the tallest formation on the Palouse, but it's not alone.

I am solid and rocky.

But I wasn't always
above for all to see.
I was once under
ground, alone, scared,
imprisoned
for years and years.
But I got pushed up
and out of the ground.
I was finally free
of the Earth's restraints.
Above the ground
for all to see.

I am Kamiak Butte.

OVER billions of years of plate tectonics, the Oceanic Plate and the North American Continental Plate helped shape the Palouse region. The Oceanic Plate subducted, or sunk beneath, the North American Plate. So much heat was created that the rock of the Continental Plate melted and formed an underground magma chamber. Over millions of years the magma chamber slowly cooled and hardened into an intrusive igneous rock called a granite batholith. This enormous mound of granite deep under the Earth's surface waited to be something even bigger and better.

I once was bubbling hot

underground as sizzling as the sun.
I was very lonely. I waited for
millions of years. Over all that time
I hardened into a HUMONGOUS mound of
crumpling crunching crashing granite.

I am a magma chamber.

FOR millions of years plate tectonics slowly forced this enormous granite mound through ten miles of the Earth's crust. The massive granite rock was pushed up and lifted more and more and more, and eventually formed the Bitterroot Mountain Range, including our very own Moscow Mountain! Millions of years of weathering and erosion from the wind and rain has broken down the exposed rock, shaping this mountain into what it is today. When you take a hike on Moscow Mountain, can you believe that you are standing on a massive, ancient magma chamber?

I am squished

underneath the ground
lonely
until one day
lonely me, a giant ball of rock,
I began to move! I got pushed
up a little every day for a few million years.
I felt the breeze.
I was crumbled and scraped.
It was painful.
Thundering.
Cracking.
After a million (or more) years
I came to tower over the land.

Seeds landed on me and grew.
Animals climbed and flew on me.
Now people
take hikes and have picnics on me.

I am Moscow Mountain!

SIXTEEN million years ago cracks in the Earth's crust, called fissures, opened up and lava gushed out. As the lava flowed and filled the landscape like a bathtub filling up, it covered many of the hills and quartzite and granite mountain peaks. But Moscow Mountain and Kamiak Butte were not completely covered. Their tips were left peeking out. The lava cooled and hardened into an extensive plain of basalt, an extrusive igneous rock. Basalt is still found all over the Palouse region today covering the mysterious peaks and valleys of the distant past.

I oozed out of the earth's cracks

I gushed,
I splattered everywhere,
like a hose, now I
cover most mountains.
Moscow Mountain
and Kamiak Butte were
bright and alive at the top.
The blooming sun shining
hot heat
on the already warm mountains.
Toasting. Over millions of years the
lava cooled and hardened.

I am a blanket of basalt that covered the
Palouse.

AFTER the basalt covered the Palouse, wind and rain eroded some of the basalt. The rain created a river that carried the basalt pieces away. As gravity pulled the river down it carved out a canyon. Today the river is in a wavy shape that looks like a snake. The Snake River has been carving at the canyon for the past sixteen million years. That's how the deep Snake River canyon was formed.

I am as cold and warm as the sun and wind

I am wavy and long
blue and green.
As I drift I trickle in the sunlight.
As I move I slice
I chisel
I sculpt my way
through the giant rock in front of me.
When I am at the end of the line
I calm down and let the people feel
my cold
smooth
rushing deep
water.

I am the Snake River.

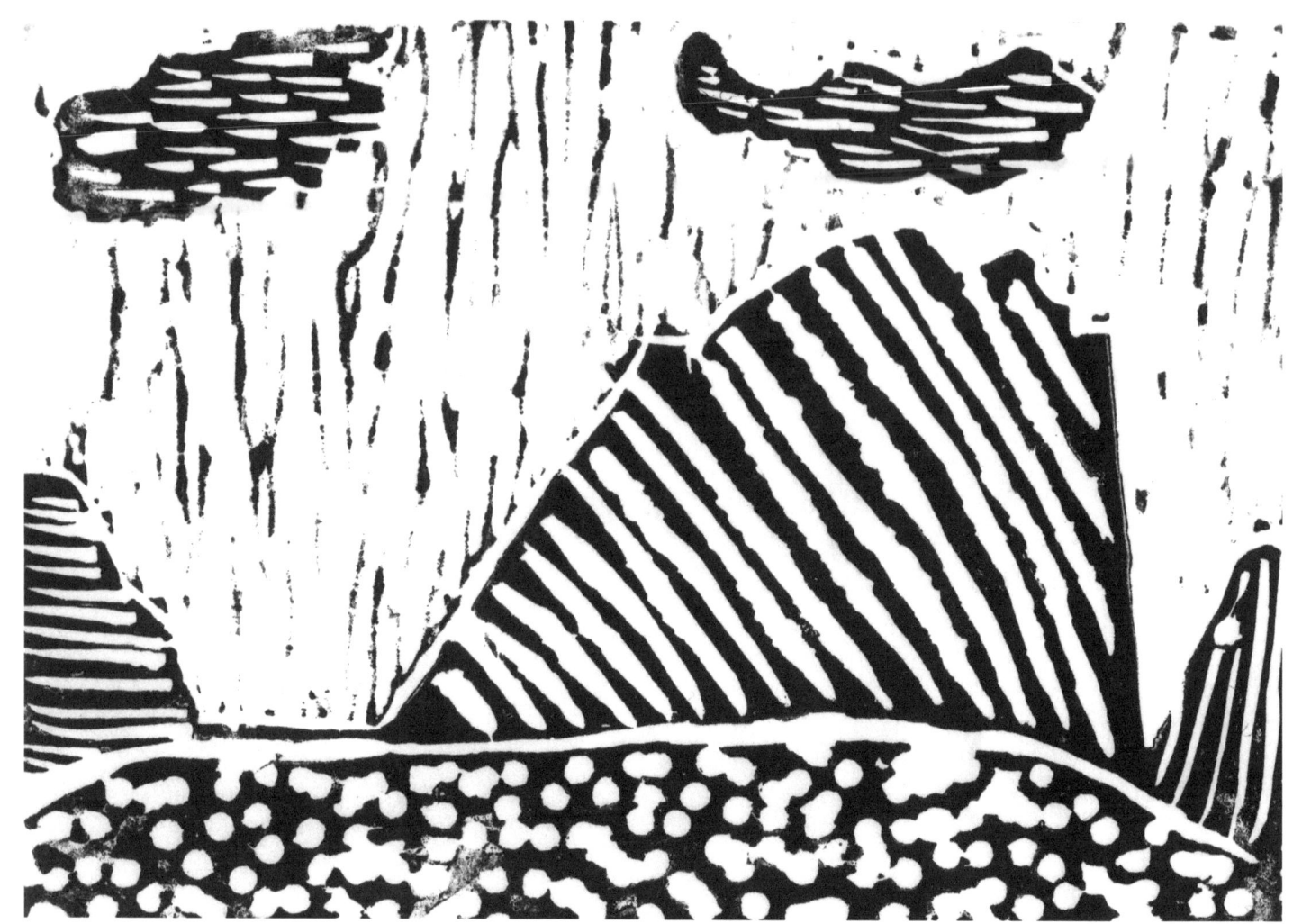

TWO million years ago glaciers slowly moved south from Canada toward the Palouse. But the glaciers didn't come all the way to the Palouse. As they scraped across the land, they carved their path and ground up the rocks beneath them. They left a trail of dirt and sediment called loess (pronounced luhs) behind them. For almost two million years, the wind picked up the loess and carried it to the Palouse, covering the basalt plains with a thick layer of dirt that would change life on the Palouse forever.

I am as soft as a rabbit's fur

Glacier's crushed me,
chopped me up
rocks into pebbles
nutritious dirt.
I was sitting in solitude
When the wind picked me up
and carried me away.
I felt like I could
touch the sky.
The wind let me go
smoothly
carefully spreading me,
as Earth's quilt.

I am loess.

FOR over a million years huge gusts of wind carried particles of sand and dirt to the Palouse. The wind shaped the loess into long, steep hills and dunes. The nutrient-rich loess holds water well, making it perfect for the farmers to grow crops on the hills. The colorful Palouse hills are blanketed with wheat, lentils, and garbanzos. It is breathtaking to know that we live in an area with something so unique and beautiful.

The loess

felt feather blowing wind softening.
I was bunching nicely, calm, piling.
I felt steep, eye catching.
I became a dune, I was moving. I felt like play
dough
in a child's hands, I have a family of friends
of dunes. I am a part of that community.
I stand tall, not alone. I am beautiful.

I am the Palouse hills.

OVER 1.5 billion years, geologic processes transformed the Palouse from a sandy ocean floor to the rolling hills and towering peaks we see today. Now the beauty of the Palouse is loved and appreciated every day by those who live here and visit the region.

Fourth grade students doing field work to discover the geologic story of the Palouse from the top of Kamiak Butte.

www.ingramcontent.com/pod-product-compliance
Lightning Source LLC
Chambersburg PA
CBHW041320180526
45172CB00004B/1166